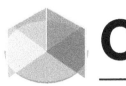

Contents

Introduction 1
Getting Started 5
Sign In, Please. 12
Penny Patterns 14
Smile 16
Pizza Pie 18
Alphabet Symmetry 20
Company Logos 22
Roadside Symmetry 24
Hex Signs 26
Snow Job 28
Circle of Friendship 30
Pinwheels 32
Symmetry Search 34
Rosette Windows 36
Driving You Crazy. 38
Name That Polygon 40
Toothpick Polygons 42
Cosmic Experience 44
Special Quadrilaterals 46
Reflection Designs 48
Circle Challenges 50
Mini-Max Perimeters 52
Perimeter Puzzles. 54
Blackline Master A: 90° and 180° Wedges. 56
Blackline Master B: 60° and 120° Wedges 57
Blackline Master C: 20°, 30°, 36°, 40°, 45°, and 72° Wedges 58
Blackline Master D: Stick Figure Wedge. 59
Blackline Master E: Grid Wedge 60
Blackline Master F: Grid Wedge Record Sheet 61
Blackline Master G: Road Signs. 62
Blackline Master H: World Wedges 63
Blackline Master I: Reflect-It™. 64
Blackline Master J: Cosmic Experience 65
Blackline Master K: Blank Circle 66
Additional Answers 67
Glossary. 70

Introduction

Today, more than ever before, teachers are looking for ways to empower students mathematically. The National Council of Teachers of Mathematics *Curriculum and Evaluation Standards for School Mathematics* stresses that *knowing* mathematics is doing mathematics. The Reflect-It™ Hinged Mirror empowers students to learn the mathematics of symmetry by doing.

What is the Reflect-It™ Hinged Mirror?

The Reflect-It Hinged Mirror has two parts: a clear, plastic base and a hinged mirror. The base, shaped like a standard protractor, has demarcations for 10 angle openings. The openings span from 10° to 180°. The two mirror faces, each 5 inches by 5 inches, are easily opened to any angle. The raised notches on the base allow the user to lock the hinged mirror into any desired setting so that reflected images can be investigated without the angle measure changing.

The angle openings on the Reflect-It base are those that are used most often throughout the student activities in this book. They are based on factors of 360° (20°, 30°, 36°, 40°, 45°, 60°, 72°, 90°, 120°, and 180°). Interestingly, each of these angles except 180° can be formed in more than one way.

Why use the Reflect-It Hinged Mirror?

Many of the activities in this book examine everyday objects, such as road signs and pinwheels, in a mathematical way. Mathematical concepts that can be explored with the Reflect-It Hinged Mirror include relationships between angle size and number of images, rotational (point) symmetry versus reflective (line) symmetry, reflections versus reproductions, and the relationship of fractional parts to one whole. The activities use an investigative approach to provide concrete experiences for discovering mathematical patterns and concepts. Students are encouraged to experiment and to test their solutions.

Many activities call for students to record their information in charts or tables, prompting students to organize their data, analyze it, and draw conclusions. Students discover that displaying their data in a spreadsheet format makes it easier to see patterns and relationships.

The Reflect-It Hinged Mirror and the activities in this book are also intended to help students achieve the five goals of mathematical literacy stressed in the NCTM *Standards*, while stimulating curiosity and providing enjoyment:

- Students will learn to value mathematics.
- Students will become confident in their ability to do mathematics.
- Students will become mathematical problem solvers.
- Students will learn to communicate mathematically.
- Students will learn to reason mathematically.

The Reflect-It Hinged Mirror has something to offer all students, regardless of their level of mathematical knowledge. Less-experienced students will be able to reach a satisfactory level of accomplishment, and more-experienced students will be stimulated to explore math concepts in greater depth.

ORGANIZATION OF THE BOOK

The ideas and activities presented in this book are a guide for working with the Reflect-It™ Mirror, not a sequential set of lessons that must be followed in order. Each activity in the book helps develop specific mathematical concepts and skills and is independent of the other activities. Each activity consists of a left-hand teacher page and a facing right-hand student page.

Teacher Pages

Each teacher page is organized into six informational categories:

Purpose–the skills or concepts that will be investigated in the activity. These include symmetry, patterns, spatial awareness, geometry, arithmetic, and algebra.

Materials–the supplies students will need to complete the activity. Materials might include recording sheets, cutout squares, paper wedges, toothpicks, pictures from magazines and newspapers, crayons or colored pencils.

Questions for Discussion–questions you may wish to pose while students work. The questions parallel the procedures on the student page and help focus on the important ideas of the activity. Many questions are open-ended and do not have unique answers. This will serve to stimulate discussion and the sharing of ideas among students.

Help for You–information about the mathematical concepts and background necessary to successfully accomplish the purpose and goals of the activity.

More–variations of the activity and applications to other subject areas for supplementary work

Answers for Student Page–possible solutions to questions on the student pages. Many activities pose questions that do not have unique solutions. In these cases, several possible answers are presented and the theory or reasoning behind the solutions is given.

Student Pages

The Student Pages can be reproduced so that each student has his or her own copy. A description of any special materials the student needs as well as the explanation of the activity is included on the student page. The student will also need the Reflect-It Mirror, including the base plate, and several sheets of paper on which to record observations and answers. You may want to encourage students to use calculators for some of the computations.

Blackline Masters

Pages 56 through 66 are eleven blackline masters that are designed to be used with specific activities. They include both patterns for blank wedges and grid wedges and special recording sheets.

USING THIS BOOK

Although each teacher brings a special, unique style of teaching to the classroom, the following suggestions may provide for a higher degree of success with the Reflect-It™ Hinged Mirror and its activities.

- Use the *Getting Started* activities (pages 7, 9, and 11) to familiarize students with the Reflect-It Hinged Mirror and how it works. These lessons were designed to be teacher-directed activities.
- Try each activity before presenting it to the class. Select activities that are challenging, yet within the range of capabilities of your students. Also, take into account the amount of time available for students to complete the activity and for a follow-up discussion.
- When you are familiar with the class dynamics and the range of individual abilities, you can choose the most appropriate ways to organize your students to work on the activities. Some suggestions include (1) teacher-directed lessons, (2) independent student work, and (3) cooperative groups in which one student manipulates the Reflect-It Hinged Mirror while the other students record the data. Vary the approaches, but provide time for students to share their results and conclusions as a whole group. This will help to strengthen your students' ability to communicate mathematically.
- Encourage students to experiment with the Reflect-It Hinged Mirror and explore on their own. They should feel free to guess, predict, theorize, and later analyze their findings. Always give students the opportunity to support their conclusions with examples.
- Use the activities as a springboard for interdisciplinary work. Emphasize the importance of mathematics in real-world situations and in different cultures and the application of mathematics to other fields of study. Suggestions for related activities that involve art, science, history, and so on, can be found on the Teacher pages under *More*.

FACILITATING STUDENTS' WORK

Act as facilitator

Facilitate students' work by discussing their strategies for working on an activity, defining new terms when necessary, encouraging a variety of approaches and solutions, and asking questions such as those suggested on the teacher pages under *Questions for Discussion*.

Monitor progress

Aid students in organizing information and setting up recording tables. When students are able to arrange their ideas in tables or charts, they help validate their own thinking processes.

Encourage self-evaluation

Help students check their answers to questions on the student pages and give them an opportunity to revise their work as they deem necessary.

Build student confidence

Encourage sharing and communication among students. Allow students to present their findings, both orally and written, and to compare their work. Display some of the students' charts, tables, or creative designs on a bulletin board for others to enjoy and reflect upon.

Getting Started

Before beginning the activities in this book, students should have the opportunity to explore the Reflect-It™ Hinged Mirror. Encourage students to experiment independently with placing small objects on the base between the faces of the mirror. Suitable objects include paper clips, coins, toothpicks, leaves, Pattern Blocks, or Attribute Blocks.

When investigating on their own, many students will change the angle opening, the number of objects placed between the faces of the mirror, or both. Suggest that students vary either the angle opening or the number of objects while leaving the other variable unchanged, so that they can begin to organize their thoughts about the effect of each individual change. Ask students to share their discoveries with their classmates.

After students have had ample time for exploration and discussion, gather them together to work on the Getting Started activities as a class.

Getting Started

Part 1

Purpose

- To become familiar with the Reflect-It™ Hinged Mirror
- To measure angles with the Reflect-It Hinged Mirror

Help for You

In the beginning, some students may need assistance positioning the Reflect-It Hinged Mirror at the correct base angle opening. Allow time for students to find the notches with the appropriate degree markings on the base. Make sure students insert the wedge tightly into the vertex of the angle opening.

More

As students explore the mirror, challenge them to answer the following questions: *How many different angles can you create? In how many different ways can each angle be made?*

Answers for Student Page

1. Answers may vary. A 90° angle can be formed by placing the mirror at 0° and 90°, 45° and 135°, or 90° and 180°.

2. Answers may vary. A 120° angle can be formed by placing the mirror at 0° and 120° or at 60° and 180°.

3. Place the mirror at 0° and 180°.

Getting Started

Part 1: How to Measure an Angle

To work with the Reflect-It Hinged Mirror, you need to know that a circle has 360°. This is a number that mathematicians agreed upon hundreds of years ago. Knowing this, you can find how many degrees there are in a half circle and a quarter circle.

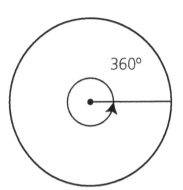

- Position the Reflect-It Hinged Mirror to form a 90° angle.
    1. Can you find two more positions you can use to make a 90° angle?

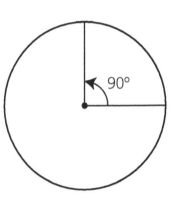

- Now position the mirror to form a 120° angle.
    2. How else can you make a 120° angle?

- The mirror can make angles up to 180°, or a half circle.
    3. How would you position the mirror to make a 180° angle?

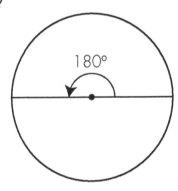

© ETA hand2mind®

7

Getting Started

Parts 2 and 3

Purpose
- To become familiar with the Reflect-It™ Hinged Mirror
- To differentiate between a reflection and a reproduction
- To investigate reflected distances

Materials

1 copy of Blackline Master D for each student, page 59

Help for You

As students examine the reflection of the stick figure in Part 2, some students may have difficulty understanding why the figure's left arm now appears on the right side. This provides an opportunity to discuss the concepts of reflection and reproduction.

The concept of reflection can be reinforced by asking two students to face each other. Ask one student to raise her or his left hand and the other student to raise her or his hand that is directly across from the raised one. The second student should identify the hand that was held up as his or her right hand.

To demonstrate the concept of reproduction, ask two students to stand side by side, each with his or her left hand raised. One student is now an exact copy or replica of the other.

In Part 3, you can reinforce the concept of reflection as students observe that the distance, as well as the triangle, is being projected across the reflecting surface.

Answers for Student Page

1. A reflection is a flip; it reverses the orientation of the image. A reproduction moves the original figure to a new position but does not alter the orientation.

2. The distance between the triangle on the *Stick Figure Wedge* and the mirror is $\frac{1}{2}$ inch. The distance between the original triangle and its image in the mirror is 1 inch, or twice as much, because there is another $\frac{1}{2}$ inch generated from the image of the triangle to the mirror.

Activities for Reflect-It™ Hinged Mirror

Getting Started

Part 2: Reflections and Reproductions

You can use a 120° angle to investigate what you can do with the Reflect-It Hinged Mirror. You will need a copy of Blackline Master. D.

- Examine the figure on the wedge. Which arm is raised?
- Position the Reflect-It Hinged Mirror to form a 120° angle. Fit the wedge into the mirror opening so that the entire figure is reflected.
- Examine the reflection of the stick figure in the left mirror face. Which arm appears to be raised now?

   1. A mirror *reflects* the original object or design. A mirror does not *reproduce*, or copy, the original object or design. Explain the difference between *reflect* and *reproduce* in your own words. Give examples.

Part 3: Reflected Distances

Set the mirror at a 120° angle. You will need your copy of Blackline Master D.

- Measure and record the distance between the drawing of the triangle and the edge of the wedge of paper.
- Estimate the distance between the original triangle and its image in the mirror.

   2. Why does the distance appear to be twice the original measurement?

Getting Started

Parts 4 and 5

Purpose
- To use symmetry to represent multiplication of groups
- To develop spatial awareness

Materials
10 small squares or 10 identical small objects

4 squares, each 1 cm × 1 cm

1 copy of Blackline Masters E and F for each student, pages 60 and 61

Help for You
In Part 4, discuss the value of organizing the information collected from activities using the Reflect-It™ Hinged Mirror in a table or chart. Allow students to suggest possible column headings. One possible chart is shown in answer 1 below.

As students record their results, they may discover the advantage of sequencing the number of groups rather than recording them randomly. This allows students to see patterns and relationships between angle measures, total numbers of groups generated, and the final multiplication sentence.

Part 5 of the activity helps develop spatial awareness with the mirror. Perimeters of the figures will vary, although the generated area remains a constant 16 square units. Help students realize that they can place the four squares on different grid squares and generate images with the same area.

Answers for Student Page

1. Students data should include:

Starting Number of Squares	Total Number of Groups Generated	Multiplication Sentence	Angle Opening in Degrees
2	2	2 × 2 = 4	180°
2	3	3 × 2 = 6	120°
2	4	4 × 2 = 8	90°
2	5	5 × 2 = 10	72°
2	6	6 × 2 = 12	60°
2	7	7 × 2 = 14	51.42°
2	8	8 × 2 = 16	45°
2	9	9 × 2 = 18	40°
2	10	10 × 2 = 20	36°

2. Start with *n* squares, where *n* can equal any number from 3 through 10. Use the same angle measures used for multiplying groups of 2 squares to write multiplication sentences about the number *n*.

3. See answers shown at left.

4. The areas of the generated images remain constant while the perimeters vary.

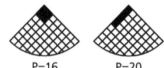

P=16   P=20
P=20   P=20
P=24   P=24
P=24   P=24

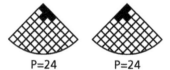

P=28   P=28

Getting Started

Part 4: Mirror Multiplications

You can use the Reflect-It Hinged Mirror to multiply whole numbers. You will need 10 small squares or 10 identical, small objects to generate a multiplication table up to 10 x 10.

- Position the mirror to form a 180° angle. Place two squares between the open mirrors to generate two groups of two squares. To find the product of 2 x 2, count the total number of squares, 4.
- Change the angle until you can see three groups of two squares. Find the product of 3 x 2.
- Continue decreasing the angle until you have generated ten groups of two squares to find the product of 10 x 2.

    1. Record your results and discuss ways of presenting them for other students to see.
    2. How can you use this technique to show the multiplication facts for 3 through 10? Demonstrate your technique to a classmate.

Part 5: Area and Perimeter

You will need a copy of Blackline Masters E and F and four one-centimeter squares.

- Position the mirror to form a 90° angle.
- Slide the mirror over the grid wedge so the wedge appears in the mirror.
- Place one square on the shaded corner square of the grid. Place the other three squares on the grid, one square per box, so that any two squares touch side to side.
- Find the area and the perimeter of the generated figure. Record them on the recording sheet and show how you placed the squares.

    3. Experiment to find as many other ways as possible to arrange the 4 squares following the rules above. Record each arrangement.
    4. What relationships did you notice among the areas and the perimeters of the generated images?

Sign In, Please

Purpose
- To explore using the Reflect-It™ Hinged Mirror
- To investigate how the angle opening affects the number of images

Materials

1 blank sheet of paper for each student
1 calculator for each student (optional)

Questions for Discussion
- Can you make your name appear seven times? Explain why it is more difficult to find the angle measure needed to do this. [51.43°]
- Are there other situations when the angle measure is not a whole number of degrees?

Help for You

As students work, they should discover several mathematical relationships. Students may find it useful to use a calculator as they work with these relationships.
- Each reflection of the angle opening is congruent to the actual angle opening.
- As the angle measure decreases, the number of images increases.
- The angle measure times the number of names equals 360°.
- 360° divided by the number of names equals the angle measure.
- If the number of names is not a factor of 360, then the angle measure will not be a whole number of degrees. If students have completed *Penny Patterns*, ask, *How do the patterns you discovered in this activity compare with the patterns you discovered in* Penny Patterns?

Answers for Student Page

1. Possible results:

Number of Names I See	Degrees In Angle Opening	Number of Names Times Degrees in Angle
2	180°	360
3	120°	360
4	90°	360
5	72°	360
6	60°	360
8	45°	360
9	40°	360
10	36°	360
12	30°	360
18	20°	360

2. Students may find the following patterns:
As the angle measure decreases, the number of images increases.
Angle measure × Number of names = 360°
360° ÷ Number of names = Angle measure.

Sign In, Please

At many state fairs or street carnivals, you can get your signature analyzed. In this activity, you get your signature multiplied.

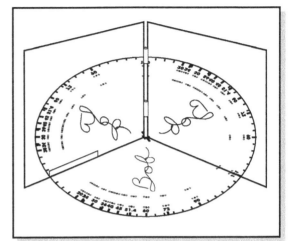

■ Write your name up or down along the center of a blank sheet of paper.

■ Position the Reflect-It Hinged Mirror at an angle of 120°.

■ Slide the sheet of paper under the base so that your signature is reflected in both faces of the mirror. You should see your signature three times—the original and two reflections.

■ Experiment to see how many different times you can make your signature appear by changing the angle openings. Find the number of degrees in each angle opening.

1. Record your results on a chart like the one below. Be sure to count the signature you wrote.

Number of Names I See	Degrees in Angle Opening	Number of Names Times Degrees in Angle

2. Make a list of all the patterns you find.

Penny Patterns

Purpose
- To find the relationship between the number of images and the angle measure
- To investigate the meaning of indirect variation

Materials
1 penny or small object for each student
1 calculator for each student (optional)

Questions for Discussion
- Was organizing your data in a chart helpful? In what way?
- What did you notice when you looked at your data?
- Describe the relationship between the angle opening in degrees and the number of pennies generated. How can you justify your answer?
- Compare this activity to *Sign In, Please*.
- Suppose you used a dot instead of a penny. How many dots would you see if the opening were 1°? 18°? Which angle opening in degrees would be needed to see 15 dots? 90 dots?

Help for You
In this activity the term *generate* is introduced. Since this term will be used throughout this book, it is important that students understand its meaning.

In this activity, the number of images increases, the measure of the angle opening decreases. This is an example of *indirect variation*. In this activity, the product of the angle opening in degrees and the corresponding number of pennies or dots generated equals 360°.

Answers for Student Page

1. Results in order from least to greatest number of images:

Number of Pennies Generated (including the original penny)	Angle Opening in Degrees	Number of Pennies Times Degrees in Angle
2	180°	360
3	120°	360
4	90°	360
5	72°	360
6	60°	360
8	45°	360
9	40°	360
10	36°	360
12	30°	360
18	20°	360

2. Patterns students may find: As the angle measure decreases, the number of images increases. Angle measure × Number of pennies generated = 360°. 360° ÷ Number of pennies generated = Angle measure. If the number of pennies generated is not a factor of 360, then the angle measure will not be a whole number.

Penny Patterns

Did you ever wish you could multiply the money you had? You can do so with your Reflect-It Hinged Mirror.

- Position the Reflect-It Hinged Mirror at any setting you wish. Place a penny or small object between the faces of the mirror. How many pennies do you see?

- Change the mirror setting. How many pennies do you see now?

- Experiment with the other mirror settings. How many different ways can you generate complete pennies? Generated pennies include the complete pennies made from the reflections and the original penny.

1. Record your results on a chart like the one below.

Number of Pennies Generated (including the original penny)	Angle Opening in Degrees	Number of Pennies Times Degrees in Angle

2. Make a list of all the patterns you find.

Smile

Purpose
- To investigate vertical line symmetry
- To investigate if a correlation exists between a person's being left-handed or right-handed and that person's photographic good side

Materials
1 full-face close-up photograph for each student, no larger than 4 in. × 4 in.

1 straightedge for each student

1 pair of scissors for each student

magazines and newspapers (optional)

It is important that each picture be taken from straight on to ensure that both sides of the face are about equal in size. If instant cameras are not available, take the photos several days before students work on this lesson to allow time to have the film processed.

Questions for Discussion
- How do the two new faces appearing in the mirror compare?
  Describe the differences and similarities.
- Which face do you like better?
- Does your face have perfect vertical line symmetry? Explain.
- Based on the results of your survey, what is the relationship between a person's photographic good side and whether that person is left-handed or right-handed?

Help for You
This activity works best as a whole-class activity with you acting as the facilitator while students collect data for the survey.

When students label the left and right sides of their face on the photograph, they must remember that their image in the photo is reversed. If students need a reminder, hold up your right hand as the students face you and ask them on which side it appears to them.

Advise students to make sure they cut the photograph through the exact center of the face to insure that each generated image will be a complete face.

Discuss the meaning of *vertical line symmetry*. A figure has vertical line symmetry if it can be folded along a vertical line so that the left half is the exact mirror image of the right half. If the figure is folded along the line of symmetry, the two halves will coincide.

More
Ask students to investigate whether animal faces have vertical line symmetry. Students can use photographs of their pets or can cut pictures out of magazines and newspapers.

Answers for Student Page
1.–3. Answers may vary, but students should notice that the two faces are not identical.

4. Possible class chart layout:

	Left Side Is Good Side	Right Side Is Good Side
Left-handed		
Right-handed		

Smile

Have you ever heard people talk about their good side or bad side when they have a photograph taken? Does your face have a good side and a bad side?

You need a close-up photograph of your full face.

■ Draw a straight line down the center of the photograph, between the eyes and through your nose and lips. Label the left side and the right side of your face on the photograph. Then cut the photograph in half, along the line.

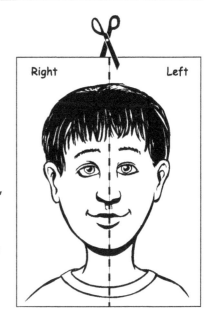

■ Position the Reflect-It Hinged Mirror to make a 120° angle. Place each half of your photograph against the sides of the mirror to make two complete faces.

1. Describe the differences between your two faces.

2. Which side of your face produces the complete image you like better? This is your good side.

3. Are you left-handed or right-handed? Is this the same as or different than the good side of your face?

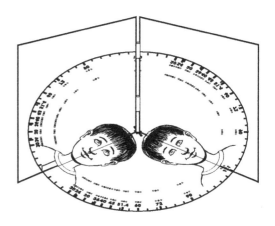

■ Combine your observations with your classmates' into a class chart.

4. What conclusions can you draw from the class chart? Write as many statements as you can.

Pizza Pie

Purpose
- To investigate the relationship between the size of a fractional part of a circle and the number of these parts in the whole circle

Materials
Cross-sections of an orange, grapefruit, lemon, lime, cucumber, or tomato, 1 per student (optional)

Questions for Discussion
- What did you notice as you positioned each pizza slice in the Reflect-It Hinged Mirror?
- What is the relationship between the number of slices in the whole pie and the portion of the pie represented by one slice?
- How could the pizza chef respond to the boy?

Help for You
The small slice is $\frac{1}{8}$ of the pizza; the larger slice is $\frac{1}{6}$. A whole pie can be generated with 8 eighths or 6 sixths.

Discuss the concept of reciprocals:
- The reciprocal of a number is 1 divided by that number, such as 6 and $\frac{1}{6}$, 8 and $\frac{1}{8}$, 10 and $\frac{1}{10}$, $\frac{1}{2}$ and 2, $\frac{3}{2}$ and $\frac{2}{3}$.
- The product of a number and its reciprocal is always 1. As the size of a slice increases, the number of slices needed to make up the whole decreases.
- As the size of a slice decreases, the number of slices needed to make up the whole increases.

More
Ask students to examine cross-sections of an orange, a grapefruit, a lemon, a lime, a cucumber, or a tomato. Ask them how they could slice each and use a slice to create a whole in the mirror. Ask students what portion of the whole they used and what size angle they used to generate each whole. Discuss what non-symmetric parts must be removed from each fruit or vegetable to maintain symmetry with the mirror images. Ask students to find other foods that can be sliced and generated in the Reflect-it Hinged Mirror.

Answers for Student Page
1. The smaller slice fits into a 45° opening, and there are 8 slices in the whole pie.
2. The larger slice fits into a 60° opening, and there are 6 slices in the whole pie.
3. The size of the whole pizza remains the same no matter how many slices the chef cuts.

Pizza Pie

Does the boy's answer make sense? You'll find out when you do this activity.

■ Place the Reflect-It Hinged Mirror on the smaller slice of pizza. Adjust the mirror until you see a whole pizza pie.

1. How many slices of pizza are in the whole pie? What is the angle opening?

■ Now place the mirror on the larger slice of pizza. Adjust the mirror until you see a whole pie.

2. How many slices of pizza are in the whole pie now? What is the angle opening this time?

3. Explain why the boy's answer in the cartoon is humorous.

© ETA hand2mind®

Alphabet Symmetry

Purpose

- To explore vertical and horizontal line symmetry
- To investigate which letters have vertical and/or horizontal line symmetry

Questions for Discussion

- What does it mean for a figure to have vertical line symmetry?
- What does it mean for a figure to have horizontal line symmetry?
- Can the symmetry of a letter be affected by the way it is written? Give an example to justify your answer.

Help for You

If necessary, review the meaning of *generating an image* with students.

A figure has line symmetry if it can be folded along a line so that the left half is the exact mirror image of the right half. If the figure is folded along the line of symmetry, the two halves will coincide.

Many students will recognize that letters may contain vertical line symmetry, horizontal line symmetry, both, or neither. It is important that students have an opportunity to experiment with the placement of the mirror on each letter to determine which symmetry, if any, exists.

When students explain vertical line symmetry or horizontal line symmetry, they may choose to use diagrams or more formal definitions.

More

Ask students to investigate other alphabets and number systems to determine the types of symmetry exhibited by their characters.

Palindromes are words or phrases that are spelled the same forward and backward. Ask students to use symmetrical letters to create palindromes that can be formed using the Reflect-It Hinged Mirror. For example, MOM can be formed with the mirror using *MC*; DAD cannot be formed with the Reflect-It Hinged Mirror.

Answers for Student Page

1.–2. Answers may vary, depending on how students write the letters. Possible results:

Vertical Line Symmetry Only	Horizontal Line Symmetry Only	Both	Neither
A, M, T, U, V, W, Y	B, C, D, E	H, I, O, X	F, G, J, K, L, N, P, Q, R, S, Z
i, v, w	c	l, o, x	a, b, d, e, f, g, h, j, k, m, n, p, q, r, s, t, u, y, z

Activities for Reflect-It™ Hinged Mirror

Alphabet Symmetry

In this activity, you will investigate letters of the alphabet. Some letters have vertical line symmetry; some have horizontal line symmetry. Some letters have both; some have neither.

The letter **A** has vertical line symmetry but not horizontal line symmetry.

How to verify vertical line symmetry

Draw a large capital **A**. Position the Reflect-it Hinged Mirror at a 180° angle. Slide it over the letter **A**, as shown on the right. If you can generate the whole letter, then the letter has vertical line symmetry. A capital letter **B** has horizontal line symmetry but not vertical line symmetry.

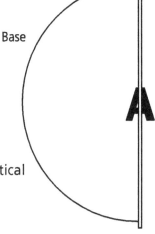

How to verify horizontal line symmetry

Draw a large capital **B**. Position the mirror at a 180° angle. Slide it over the letter **B**, as shown on the right. If you generate the whole letter, then the letter has horizontal line symmetry.

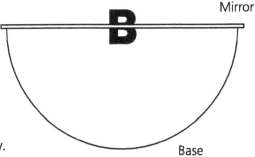

1. Predict which upper-case (capital) letters have line symmetry. Test each letter with the mirror. Record your results on a chart like the one below.

Vertical Line Symmetry Only	Horizontal Line Symmetry Only	Both	Neither
A	B		

2. Predict which lower-case letters have line symmetry. Then test each letter with the mirror. Record your results on your chart.

© ETA hand2mind®

Company Logos

Purpose
- To investigate vertical, horizontal, and diagonal line symmetry in logos

Materials
magazines, newspapers, and product packaging for the class to share
1 pair of scissors for each student

Questions for Discussion
- What are some different ways you can use the Reflect-It Hinged Mirror at the 180° angle opening to generate the former Citibank logo?
- How could you change the logo so that it has diagonal symmetry?
- What parts of a logo might prevent it from having line symmetry? Give examples as you explain your answer.

Help for You
Students may discover more than one way to reproduce the logo design with the same angle opening. This affords an opportunity to discuss three different types of line symmetry—vertical, horizontal, and diagonal.

As students discuss different types of line symmetry, include logos that do not have line symmetry. Students should be able to identify the non-symmetric elements and suggest how the logo could be changed so that it would have line symmetry.

More
Ask students to create three different, original logos: one that has only vertical line symmetry; one that has only horizontal line symmetry; and one that has vertical, horizontal, and diagonal symmetry. Then have students use the Reflect-It Hinged Mirror to check their designs.

Answers for Student Page
1. Answers may vary. Students' answers should reflect an understanding of the three types of line symmetry.
2. The former Citibank logo has horizontal and vertical line symmetry.
3. Answers may vary. Students' answers should reflect an understanding of the three types of line symmetry.
4. Answers may vary, but they should reflect an understanding of line symmetry.

Activities for Reflect-It™ Hinged Mirror

Company Logos

A *logo* is a company's design or symbol. Companies such as TV stations, sneaker manufacturers, and restaurants use logos on their products or in their advertising. A logo helps people remember a company.

This once was part of Citibank's logo*.

1. Use the Reflect-It Hinged Mirror to find different ways to generate the whole logo. Describe what you found.

■ Position the mirror to form a 180° angle. Slide the mirror across this logo to discover which type(s) of line symmetry it has.

2. What kind of symmetry does this logo have?

■ Cut out at least 10 different logos from newspapers, magazines, product packaging, and advertising flyers.

■ Experiment with the logos. What angle openings generate each logo? Use the mirror to find the type(s) of line symmetry each logo has.

3. Record your results on a chart like the one shown below.

Logo	Angle Openings that Generate the Logo	Types of Line Symmetry

4. How are the logos alike? How are they different? Write as many statements as you can.

*The Compass Device logo is a registered service mark of Citibank, N.A., and is used with permission.

© ETA hand2mind®

Roadside Symmetry

Purpose
■ To investigate vertical, horizontal, and diagonal line symmetry in road signs

Materials
1 copy of Blackline Master G for each student, page 62

Questions for Discussion
■ Which road signs on the blackline master have no line symmetry?

■ Why doesn't the stop sign have line symmetry?

■ What conclusion can you draw about the types of line symmetry in the signs that can be generated from the 90° angle opening of the mirror? Use the data in your chart.

Help for You
Students may discover more than one way to reproduce the road signs by using the same angle opening. This affords an opportunity to discuss three different types of line symmetry—vertical, horizontal, and diagonal.

Students may discover that if the image generated using the 90° angle opening duplicates the original sign, then the sign possesses vertical and horizontal line symmetry.

As students discuss their observations, include other traffic signs they may have seen. Ask them to consider the design on the sign, its shape, the color placement, and any wording that there may be. For example, if there were no text on the stop sign, it would have vertical, horizontal, and diagonal symmetry.

More
Ask students to design their own symmetric signs for specific road conditions or for landmarks.

Answers for Student Page
1. Answers may vary. **2.–3.** Possible results are shown in the chart below:

Sign	Angle Openings That Generate the Sign	Types of Line Symmetry	Can you generate the sign from a 90° angle opening?
Hospital	90°, 180°	vertical, horizontal	yes
Side Road	180°	horizontal	no
Road Narrows	90°, 180°	vertical, horizontal	yes
Campground	180°	vertical	no
Road Hazard		none	no
Traffic Circle		none	no
End of Road	90°, 180°	vertical, horizontal	yes
Yield	60°, 120°, 180°	vertical	no
Cross Road	45°, 90°, 180°	vertical, horizontal, diagonal	yes
Two-Way Traffic		none	no
Stop Ahead	180°	vertical	no
Stop		none	no

Roadside Symmetry

Road signs give us different kinds of information. For example, this sign indicates that a hospital is located nearby.

You will need a copy of Blackline Master G.

1. Use your Reflect-It Hinged Mirror to find different ways to generate the whole hospital sign.
   Describe what you found.

▪ Position the Reflect-It Hinged Mirror to form a 180° angle. Slide the mirror across the sign to discover which type(s) of line symmetry it has.

2. What kind of symmetry does the hospital sign have? Record your observations on a chart like the one below.

▪ Move the mirror to form a 90° angle. Slide the mirror across the hospital sign until $\frac{3}{4}$ of the sign is covered. Is the entire sign generated by the uncovered portion of the sign and its reflection?

▪ Experiment with the other signs. What angle openings generate each sign? Use your mirror to find the type(s) of line symmetry each sign has.

3. Record your results.

Sign	Angle Openings That Generate the Sign	Types of Line Symmetry	Can you generate the sign from a 90° angle opening?

© ETA hand2mind®

Hex Signs

Purpose

To investigate symmetry in Pennsylvania Dutch hex signs

Materials

1 copy of Blackline Masters B and K for each student, pages 57 and 66

1 blank circle for each student

1 each blue, red, green, yellow, and black markers or crayons for each student

atlases, history books, and/or old books (optional)

Questions for Discussion

- Which Reflect-It Hinged Mirror angle openings can be used to generate hex sign **A**? What is the smallest angle opening?
- Which mirror angle openings can be used to generate hex sign **B**? What is the smallest angle opening?
- How could color affect the symmetry of each hex sign?

Help for You

You may wish to ask students to find pictures of other hex signs with hearts, flowers, leaves, birds, or unicorns. Encourage discussion about their symmetric elements, as well as the impact of the colors used. Encourage students to find if hex signs are still used today and for what purpose. (You may also wish to point out that in this context, *hex* does not mean *six*.)

When students place their own design in the mirror, they will generate a totally symmetric hex sign. Ask students to devise a color scheme so that the symmetry is no longer preserved.

More

All maps have directional compass roses to indicate the cardinal directions—north, south, east, and west. Have students look at a variety of compass roses on maps in atlases or history books. Ask students if it would be possible to create a compass rose using their mirrors. Which mirror angle openings could be used to generate the compass roses? Must the north, south, east, and west labels be omitted to have symmetry? Have students design compass roses using the 45° or 90° angle openings on the mirror.

Answers for Student Page

1. hex sign **A**: 45°, 90°, 180°
   hex sign **B**: 60°, 120°, 180°

2. hex sign **A**: 45°
   hex sign **B**: 60°

Hex Signs

Throughout southeastern Pennsylvania in Pennsylvania Dutch country, brightly colored circular disks with geometric and floral designs adorn barns and stables. In the past, farmers thought that these disks, or hex signs, could protect their animals from diseases and ward off evil spirits.

Here are two examples of hex signs.

- Experiment with your Reflect-It Hinged Mirror to discover the different

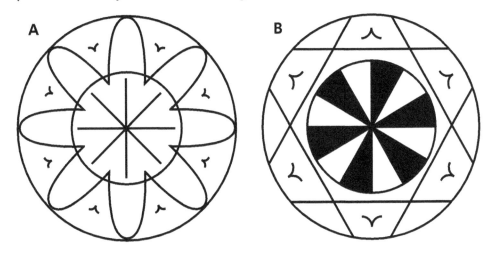

  angle measures or patterns that will generate hex sign **A**.

  1. Make a list or draw a sketch to record your observations.

  2. What is the smallest angle opening that generates the complete hex sign?

- Repeat the above activity with hex sign **B**.

Now create your own hex sign. You will need one copy each of Blackline Masters B and K.

- Draw a design on the paper wedge on Blackline Master B.
- Position your design in the mirror to generate a hex sign.
- Copy the generated hex sign onto the blank circle on Blackline Master K.
- Decorate your hex sign using bright colors.

Snow Job

Purpose
- To investigate symmetry in snowflakes

Materials

1 copy of Blackline Masters A–C for each student, pages 56–58

Questions for Discussion

- What does the word *plate* suggest? What does the word *stellar* suggest? Why do you think snowflakes are called "plate" and "stellar"? (Stellar snowflakes resemble six-pointed stars; plate snowflakes are like a hexagonal plate.)
- Which angle openings generate the original snowflakes? Which angle opening is used for line symmetry on the snowflake?
- How can you place the Reflect-It Hinged Mirror on a snowflake to generate an identical snowflake?
- For each mirror setting, in how many different positions can you place the mirror to generate each snowflake?

Help for You

Have students work with partners or in small groups. They can share their discoveries with other students and have several examples to look at simultaneously.

More

Ask students to investigate other crystalline shapes and solid precipitation such as hail and sleet. Can any of these forms be generated using their mirrors? Have the students record their findings on blank wedges or on a sheet of paper.

Answers for Student Page

1.–2. Students should recognize that these mirror settings and angle openings can be used to generate both snowflakes: 30°, 60°, 120°, 180°. Wedge designs that will generate each snowflake:

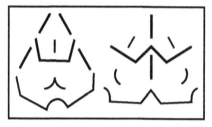

Plate Snowflake   *Stellar Snowflake*

3. Answers may vary.
4. Answers may vary.

Snow Job

All snowflakes are different, but most are six-sided geometric shapes.

Here are two examples of snowflakes.

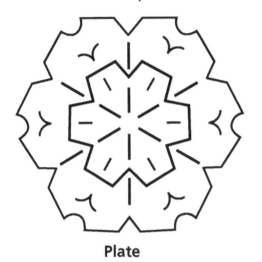

Plate   Stellar

- Place the Reflect-It Hinged Mirror on the plate snowflake. Experiment to find which mirror settings will generate the original snowflake. Find all the positions for each mirror setting.

- Repeat with the stellar snowflake.

    1. Record your observations.

    2. What patterns do you notice? Write as many statements as you can.

- Design your own snowflake. Select any mirror setting you wish. Use your mirror and the appropriate wedge to create a design that will generate your snowflake.

    3. Record your design.

    4. On another wedge, generate your same snowflake in a different way.

© ETA hand2mind®

Circle of Friendship

Purpose
- To determine the original picture needed to generate a final image
- To visualize the difference between a flip (or reflection) and a reproduction (or duplication)

Materials
1 copy of Blackline Master H for each student, page 63

Questions for Discussion
- What does the terms *flip* or *reflect* suggest when you use a mirrored surface?
- How can you explain the meaning of the terms *reproduce* and *duplicate* in connection with a mirror?
- How many girls and boys should you draw on the 90° world wedge if you want four of each in the final image?
- How many girls and boys should you draw on the 45° world wedge if you want four of each in the final image? if you want eight of each in the final image?
- How do the water and landmasses look when the world wedge is in the mirror?

Help for You
The mirrored surface will flip, or reflect, the wedge; the mirror will not reproduce, or duplicate, it. When students place the 90° world wedge with the entire boy and girl drawing in the mirror, the generated image shows an alternating pattern of two boys, two girls, two boys, and two girls instead of the original alternating pattern of one boy, one girl.

When students place the world wedge in the mirror, the water and landmasses will look distorted because the same design appears on each sector in the image. Tell students to focus only on the arrangement of the boys and girls.

More
Ask students to experiment with changing the number of boys and girls, their arrangement, or the angle opening. For example, can students generate an image of boy, boy, boy, girl, girl, girl on the global surface using a 180° angle opening?

Answers for Student Page
1. The mirrored surface will flip, or reflect, the wedge; the mirror will not reproduce, or duplicate, it.

2.

3.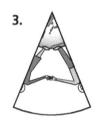

Activities for Reflect-It™ Hinged Mirror

Circle of Friendship

Pictures like this are often used to represent world friendship. What do you notice about the arrangement of boys and girls?

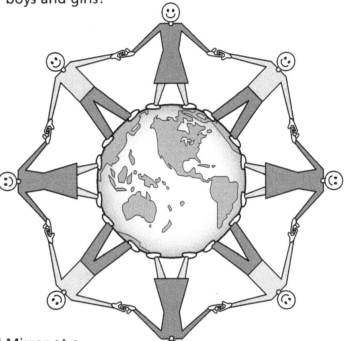

■ Position the Reflect-It Hinged Mirror at a 90° angle and place it on the wedge on the right.

   1. What do you notice about the arrangement of boys and girls in the generated image? Why do you think this happens?

■ Use a blank 90° world wedge in the mirror.
  Draw an arrangement that will generate the original picture.
  Use pencil so you can change your drawing if necessary.

   2. Compare drawings with a classmate. How are the drawings alike? How are they different?

■ Position your mirror at a 45° angle.

   3. On a blank 45° world wedge draw an arrangement that will generate the original picture.

© ETA hand2mind®

Pinwheels

Purpose
- To investigate rotational symmetry
- To determine whether all designs with rotational symmetry also have reflective symmetry

Materials
1 copy of Blackline Master A for each student, page 56
1 cardboard circle for each student (optional)
1 short pencil for each student (optional)

Questions for Discussion
- How does the word pinwheel suggest rotational symmetry?
- Each 90° wedge represents a quarter of the pinwheel. Which quarter design generated the original pinwheel? Which quarter design did not?
- Is there more than one way to draw a design on the 90° wedge for the pinwheel that can be generated with the Reflect-It Hinged Mirror? What are they?
- What do you notice when you compare the quarter designs for pinwheels **A** and **B**?

Help for You
A figure has rotational symmetry if it can be rotated around a pivot point through an angle between 0° and 360° to generate an image identical to the original image.

When students compare the designs for pinwheels **A** and **B**, they should notice that pinwheel **B** has reflective symmetry, including vertical, horizontal, and diagonal line symmetry. Pinwheel **A** does not have these symmetries. This reflective property is the reason why only pinwheel **B** can be generated with the mirror, while pinwheel **A** cannot.

The 90° wedge drawings that students create for pinwheel **B** also have line symmetry with respect to a line of symmetry through the center of the wedge. The wedge drawing for pinwheel **A** does not.

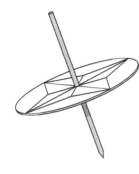

More
Have each student copy one of the pinwheels onto a 3-inch cardboard circle. Punch a hole in the center of the cardboard circle. Insert a short pencil through the hole so that approximately half an inch protrudes below the disk.

As students spin these wheels, sometimes the pattern will look like it is spinning backward, look like it is spinning forward, or look like it is not moving at all. Ask students to explain why this happens. Have them research what the stroboscopic effect is.

Answers for Student Page

1. These designs will generate pinwheel **B**.

2. There is no wedge design that will generate the original image of pinwheel **A** because pinwheel **A** does not have reflective symmetry.

Pinwheels

Each of these pinwheels has *rotational symmetry*. Rotational symmetry means that you can generate the original picture by rotating or pivoting it around its center.

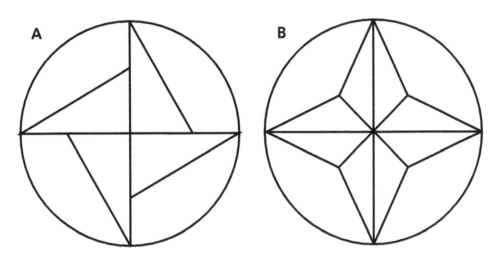

- Only one of the pinwheels can be generated with the Reflect-It Hinged Mirror. Experiment to find out which one. Position the mirror to form a 90° angle. Use blank 90° wedges.

    1. Record all the designs that will generate the pinwheel that can be generated with your mirror.

    2. A design that can be generated with your mirror has reflective symmetry. Explain why it is impossible to generate both pinwheels with your mirror.

    3. Draw your own pinwheels with rotational symmetry. Make one that can be generated with your mirror and one that cannot be generated with your mirror.

Symmetry Search

Purpose
- To investigate vertical, horizontal, and diagonal line symmetry
- To form mental images of an object's line(s) of symmetry

Materials
A variety of magazines and newspapers

Questions for Discussion
- What is vertical line symmetry? horizontal line symmetry? diagonal line symmetry? Give an example of each.
- How did you find the folding line(s) for each picture?
- Which type of symmetry did you observe most often in your pictures? Are there elements in each picture that prevent it from having total symmetry? Explain.

Help for You
You may wish to review the concept of line of symmetry as students work on this activity. Explain that each folding line is a line of symmetry.

Having students sort their pictures according to different types of symmetry provides an opportunity a class discussion. Encourage students who are not certain about where to fold or place a picture to ask classmates for their help.

More
Students' collages might remind them of a kaleidoscope. You might wish to explain that a kaleidoscope contains two mirrors, usually set at a 60° angle. The movement of small, loose pieces of colored glass between the two mirrors produces an infinite variety of patterns.

Answers for Student Page
1. Answers may vary.
2. Answers may vary.

Symmetry Search

A collage is a work of art made by pasting pictures onto a sheet of construction paper or poster board. You can use the Reflect-It Hinged Mirror to make an creative collage.

- Carefully cut out from magazines, newspapers, etc. 20 pictures of objects that you think might have vertical, horizontal, or diagonal line symmetry. Number the pictures from 1 to 20.
- Sort the pictures into piles based upon the type(s) of line symmetry you think each has: only vertical; only horizontal; only diagonal; vertical and horizontal; all three types; or no symmetry.
- Fold each picture along what you think is a line of symmetry. Remember, some pictures may have more than one line of symmetry. Check all of them with your mirror. If you can generate the whole picture with the mirror, then it has line symmetry.

1. Record your observations in a chart like the one below.

Picture	Vertical Line Symmetry Only	Horizontal Line Symmetry Only	Diagonal Line Symmetry Only	Vertical and Horizontal Line Symmetry	Vertical, Horizontal, and Diagonal Line Symmetry	No Symmetry

- Position your mirror to form a 120° angle. Place several of the folded pictures against the two edges of the mirror to create a symmetric magazine collage.

2. Which pictures work well in your collage? Which pictures do not work well? Write a statement explaining why this is so.

- Share your collage with your classmates.

Rosette Windows

Purpose
- To investigate the role of rotational and reflective symmetry in rosette windows

Materials
books about stained glass, churches, or cathedrals
1 copy of Blackline Masters A–C for each student, pages 56–58
1 box of colored pencils or crayons for each student
black construction paper (optional)
pieces of colored cellophane (optional)
pictures of flowers (optional)

Questions for Discussion
- What are the main colors in the rosette windows you found? What is the background color?
- What parts of the design do not preserve the symmetry of the rosette window? Show examples.
- How many petals are in each design?
- If you know the number of petals in the design, how can you find the angle measure of one petal?

Help for You
The angle measure of one petal in a rosette window design can be found by dividing 360° by the number of petals.

Students may wish to continue their design and make the entire rosette window out of heavy black construction paper and colored cellophane. Finished rosettes can be hung on the windows of the classroom.

More
Many flowers exhibit reflective symmetry and their images can be generated in the mirror. Ask students to bring in pictures of flowers whose likeness can be generated using a single petal and the mirror.

Answers for Student Page
1. Answers may vary.
2. Answers may vary.

Rosette Windows

Many churches and cathedrals are decorated with magnificent stained glass windows. Some of these windows are round and contain what look like flower petals. These are called *rosette windows*.

To the right are some examples of rosette windows.

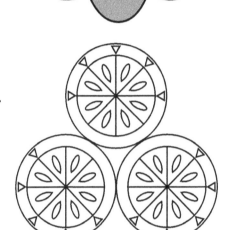

■ Look in books or search the Internet (possible search words: stained glass, churches, or cathedrals) to find pictures of rosette windows.

   1. Make a chart like the one below. For each rosette window, find and record the information indicated.
   2. What did you notice about the rosette window designs? Write all the statements you can.

■ Now create your own rosette window. Choose any angle opening. You will need blank paper wedges for that size of angle opening. You will also need colored markers or crayons.

Where Window Is Located	Number of Petals in Design	Glass Colors	Background Color	Nonsymmetric Items of Colors	Angle Opening That Will Generate One Petal

© ETA hand2mind®

Driving You Crazy

Purpose
- To examine line symmetry and rotational symmetry in wheel and hubcap designs

Materials
assorted magazines, newspapers, or car and truck brochures
1 pair of scissors for each student
2–3 pieces of tracing paper for each student

Questions for Discussion
- What non-symmetric elements did you exclude when you considered the symmetry in your pictures of wheels and hubcaps? Give examples.
- How did the number of wheels with line symmetry compare to the number of wheels with rotational symmetry? Can you find wheels with rotational symmetry that also have line symmetry?
- What was the most common number of petals or patterns on the wheels?
- What angle measures are needed to generate one pattern on the wheel?
- Did you find any wheels that did not have any type of symmetry? Give examples.

Help for You

If necessary, review the meaning of rotational symmetry. (A figure has rotational symmetry if, after a specific number of rotations of less than 360° around an internal pivotal point, the figure returns to its original position.)

Students can usually easily identify non-symmetric elements on the wheels, such as trademarks, lug nuts, tire stems, and even dents. The logos of many car manufacturers can be used as additional examples of symmetry.

Students can find the angle measure of one pattern by dividing 360° by the number of patterns, petals, or star points. For example, suppose a wheel has 5 patterns, the most common found. The angle measure of one pattern is 360° ÷ 5, or 72°.

To test if a wheel has line symmetry, students can fold the picture on what they think is a line of symmetry and place the fold against one side of their mirrors. If the reflection is the same as the original wheel, the picture has line symmetry.

To test if a wheel has rotational symmetry, students can trace the outline of one petal or pattern. If the traced pattern fits exactly over each of the other petals, the picture has rotational symmetry.

More

Students can find pictures of other examples of rotational symmetry, such as sunbursts, jewelry, flowers, or light fixtures. Have students create a bulletin board display of their pictures.

Answers for Student Page

1.–3. Answers may vary, but should reflect student's understanding of line and rotational symmetry.

Driving You Crazy

Car and truck companies keep coming up with new wheel and hubcap designs. Their patterns usually have some type of symmetry. The one most commonly used is rotational symmetry.

■ Cut out at least six pictures of wheels or hubcaps from magazines, newspapers, or car and truck brochures.

■ Use your Reflect-It Hinged Mirror to decide if each wheel or hubcap design has line symmetry or rotational symmetry. Focus only on the wheel or hubcap design. Ignore non-symmetric items, such as trademarks, bolts, and tire stems in your decision.

1. Record your observations for each wheel or hubcap on a chart like the one below.

Car Make and Model	Year	Number of Petals, Patterns, or Star Points in Wheel or Hubcap Design	Type(s) of Symmetry	Angle Opening for One Petal, Pattern, or Star Point

2. Create your own wheel or hubcap design. Decide if it will have line symmetry and/or rotational symmetry.

3. Add information about your design to your chart.

Name That Polygon

Purpose
- To use the Reflect-It Hinged Mirror to generate regular polygons
- To find the measure of each interior angle of a regular polygon

Materials
1 copy of Blackline Master I for each student, page 64

Questions for Discussion
- What is a polygon? Draw or describe examples.
- What did you notice about the sides of the triangle generated? about the angles? What was the measure of each base angle of the triangle?
- Was the chart helpful in organizing your information? In what way?
- What is the relationship between the angle opening on the mirror and the number of sides in the reflected polygon? Justify your answer.
- How did you find the measure of an interior angle of a polygon?

Help for You
Most students should be familiar with the terms *polygon*, *regular polygon*, *congruent*, *equilateral*, and *equiangular*.

- A polygon is equilateral if all its sides are congruent.
- A polygon is equiangular if all its angles are congruent.
- A regular polygon is equilateral and equiangular.

To find the measure of an interior angle in a regular polygon, students can look at the triangle formed by the sides of the mirror and the line segment. The measure of the Reflect-It Hinged Mirror angle opening is the same as the measure of the triangle's vertex angle. Students will need to know that the sum of the angle measures of a triangle is 180° and that the triangle formed by the line segment and the sides of the mirror is isosceles—it has two sides equal in length and the two angles opposite these sides are equal in measure.

As students record their results, they may see the advantage of sequencing the angle openings from greatest to least or least to greatest, rather than recording them randomly. Sequencing makes it easier to find patterns such as:

As the angle opening decreases, the number of sides in the regular polygon increases.

The product of the angle opening in degrees and the number of sides in a regular polygon equals 360°.

Answers for Student Page
1. See page 67 for answers.
2. When a triangle is formed with the sides of the mirror and the line segment, the measure of the mirror angle opening is the same as the measure of the triangle's vertex angle.

Name That Polygon

A *polygon* is a simple closed figure made of line segments. If all the angles of a polygon are equal in measure and all the sides are equal in length, then the polygon is a *regular polygon*. You can use the Reflect-It Hinged Mirror to generate many different regular polygons.

You will need a copy of Blackline Master I.

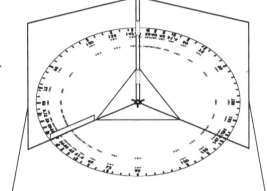

■ Place the base of the Reflect-It Hinged Mirror on top of the diagram of the Reflect-It on Blackline Master I. Make sure that the center point on the base lines up with the center point on the diagram.

■ Position the mirror 30° and 150° to form a 120° angle. You should generate a triangle.

■ Experiment with other mirror settings to discover all the polygons you can generate.

1. Record your findings on a chart like the one below.

Reflect-It Setting	Angle Opening in Degrees	Number of Sides in Polygon	Name of Polygon	Measure of Each Angle in the Polygon

2. Explain how you can use your mirror to find the measure of an interior angle of a regular polygon.

© ETA hand2mind®

Toothpick Polygons

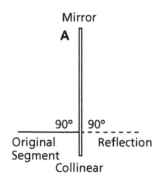

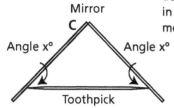

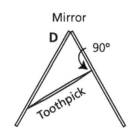

Purpose
- To use the Reflect-It Hinged Mirror to generate polygons using a segment and its reflection as one side of the polygon

Materials
1 toothpick, preferably with flat sides, for each student

Questions for Discussion
- What figures were you able to generate with the toothpick using a 60° angle opening?
- How did you position the toothpick to form an equilateral triangle using a Mirror 60° opening?
- What figures were you able to generate with the toothpick using a 45° angle opening?
- How did you position the toothpick to form a square using a 45° opening?

Help for You
When a line segment is reflected in a mirror, the original segment and its reflection can be collinear; that is, the reflection will look like a continuation of the original line (or toothpick, as in this activity) only if the original segment is placed perpendicular to the mirror. (See illustration *A*.) If the original segment is not perpendicular to the mirror, the two line segments will share a common end point but they will not be collinear. (See illustration *B*.)

As students experiment with the placement of the toothpick between the faces of the mirror, they should discover that they can generate different figures. Each generated figure depends upon the angle that the toothpick makes with the sides of the mirror. For example, in a 60° angle, the toothpick can produce hexagons with or without angles of the same measure, an equilateral triangle, or star-like shapes with three points.

Students should be aware of two special cases:
- If the angles formed by the toothpick and the sides of the mirror are the same, the generated figure will be equilateral. (See illustration *C*.)
- If the angle formed by the toothpick and one mirror is 90°, the generated figure will again be equilateral, but it will have half the number of sides as the number of generated segments. (See illustration *D*.)

Answers for Student Page
1.–3. Answers may vary.

4. The toothpick must be positioned perpendicular to one of the sides of the mirror.

Toothpick Polygons

In this activity, you will investigate regular polygons. All the angles of a regular polygon are equal in measure and all the sides are equal in length.

You will need a toothpick for this activity.

- Position the Reflect-It Hinged Mirror to form a 60° angle.

- Place the toothpick between the faces of the mirror. Slide the toothpick until you generate an equilateral triangle.

   1. Describe your results or make a sketch to show how you placed the toothpick.

- Reposition your mirror to form a 45° angle.

- Place the toothpick between the mirrors. Slide it until you generate a square.

   2. Describe your results or make a sketch to show how you positioned the toothpick.

- Experiment with different mirror settings to find other regular polygons you can generate.

   3. Describe your results or make a sketch to show how you positioned the toothpick to generate each polygon. Which mirror setting did you use for each polygon?

   4. What is the relationship between the toothpick and the faces of the mirror?

Cosmic Experience

Purpose
- To generate stars using the Reflect-It Hinged Mirror
- To examine the relationship between the mirror setting and the number of points on a star

Materials
1 copy of Blackline Master J for each student, page 65

Questions for Discussion
- How many points are on the star?
- What is the angle opening in degrees?
- Express the relationship between the angle opening in degrees and the number of points in the generated star. How can you test if your relationship is true for all the stars?
- What angle opening would you use to generate a star with 4 points? with 40 points?
- How many points would a generated star have if the angle opening were 5°? 15°?
- How do the two stars generated with wedge W compare? Which do you like better? Why?

Help for You
If students choose their wedges in an order based on either decreasing or increasing angle measure, they will find it easier to observe patterns such as the following:
- As the angle measure of the wedge increases, the number of points on the reflected star decreases.
- The angle measure times the number of points on the generated star equals 360°.

Although wedge W uses the same angle opening to generate two stars with the same number of points, the stars create two different visual images. The deeper the V is cut into the wedge, the sharper the points on the star become.

Answers for Student Page
1. Answers depend on which wedge students choose.
2. Results:

Wedge	Number of Points on Star	Angle Opening in Degrees
R	12	30°
T	9	40°
Q	8	45°
S	6	60°
P	5	72°
W	10	36°

3. It has two stars, one within another.
4. See chart above.
5. Both stars have the same number of points. The points of the inner star are deeper than the points of the outer star.

Cosmic Experience

One place to find stars is in the sky. In this activity, you will investigate how to create stars using the Reflect-It Hinged Mirror.

You will need a copy of Blackline Master J.

■ Start with wedges *P*, *Q*, *R*, *S*, and *T*. Choose one wedge and place it in the Reflect-It Hinged Mirror. Move the mirrors until you generate a star.

1. How many points does the star have? What mirror setting did you use?

■ Repeat with the other four wedges.

2. Record your observations on a chart like the one below.

Wedge	Number of Points on Star	Angle Opening in Degrees

■ Now examine wedge *W*. It has two *V*s. Place wedge *W* in the mirror. Move the mirror until you generate two stars.

3. What did you notice? How are these stars different from the others?

4. Record your observations about wedge *W* in your chart.

5. Compare the inner star to the outer star. Write as many statements as you can.

Special Quadrilaterals

Purpose
- To investigate the properties of special quadrilaterals

Materials
2 sheets of grid paper for each student

Questions for Discussion
- What is a parallelogram? a rectangle? a rhombus? a square?

The following questions can be asked for each quadrilateral generated:
- What did you notice about the length of the sides in the figure?
- How could you find the length of each side of the figure?
- How could you find the length of each diagonal in the figure?
- What is the relationship between the two diagonals in the figure?
- What did you notice about the opposite angles in the figure?
- How do the diagonals relate to the angles?

Help for You

You may wish to review the properties of a square and a rhombus with the students. Some students may identify the rhombus as a diamond.

Square	Rhombus
equilateral	equilateral
equiangular	opposite angles are congruent
all angles are right angles	opposite sides are parallel
opposite sides are parallel	diagonals bisect opposite angles
diagonals bisect opposite angles	diagonals bisect each other
diagonals are congruent	diagonals are perpendicular
diagonals bisect each other	
diagonals are perpendicular	

Encourage students to predict the type of quadrilateral they will generate before they place their mirrors on each diagram.

To determine the lengths of the sides in each quadrilateral, students can estimate, measure, or apply the Pythagorean theorem ($a^2 + b^2 = c^2$, where a and b represent the lengths of the legs of a right triangle and c represents the length of the hypotenuse).

Answers for Student Page

1. square

2. All sides are congruent; all angles are congruent; and the diagonals are congruent and perpendicular.

3. rhombus or parallelogram

4. All sides are congruent; opposite angles are congruent; and the diagonals are perpendicular.

5. Answers may vary.

Special Quadrilaterals

Parallelograms, rectangles, rhombi, and squares are quadrilaterals whose sides, angles, and diagonals have special properties. You can use the Reflect-It Hinged Mirror to investigate some of these properties.

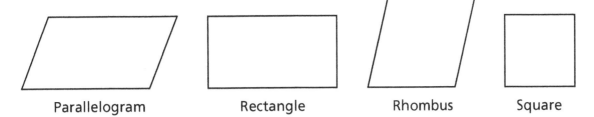

Parallelogram   Rectangle   Rhombus   Square

■ Position the Reflect-It Hinged Mirror on diagram A to form a 90° angle.

■ Slide your mirror over the grid wedge so it appears in the mirror, generating a quadrilateral and its diagonals.

   1. What quadrilateral did you generate?

   2. Describe what you notice about the sides, the angles, and the diagonals of this quadrilateral.

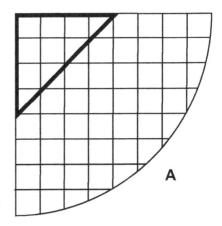

■ Now repeat with diagram B. Generate another quadrilateral and its diagonals.

   3. What quadrilateral did you generate this time?

   4. Describe what you notice about the sides, the angles, and the diagonals of this quadrilateral.

   5. On a separate sheet of grid paper, draw other diagrams to generate different-sized special quadrilaterals.

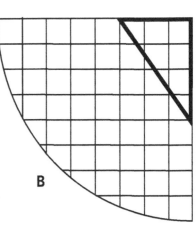

© ETA hand2mind®

Reflection Designs

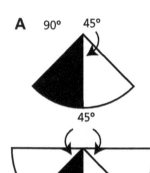

A

Purpose
- To explore the relationship between a pattern and its reflection

Materials
several copies of Blackline Masters A–C for each student, pages 56–58
1 crayon or marker for each student

Questions for Discussion
- How many ways did you find to generate each design?
- What is the relationship between the number of sections in a design and the angle opening that generates the design?

Help for You
Students will observe that the mirror reflects, or flips, the design they draw on each wedge. Encourage students to brainstorm the relationship between the number of sections in each design and the angle opening used to generate it. Guide students by asking them to find the factors of the number of sections in each pattern. Ask how these factors would relate to the angle openings.

Help students discover the following relationships:

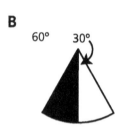

B

4 sections	6 sections	8 sections
factors of 4: 2, 4	factors of 6: 2, 3, 6	factors of 8: 2, 4, 8
angle openings: 90°, 180°	angle openings: 60°, 120°, 180°	angle openings: 45°, 90°, 180°
4 × 90° = 360°	6 × 60° = 360°	8 × 45° = 360°
2 × 180° = 360°	3 × 120° = 360°	4 × 90° = 360°
	2 × 180° = 360°	2 × 180° = 360°

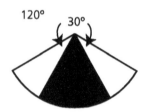

Students may discover different ways to color the wedges to generate the same images. Help students conclude that the designs would be the same, but the color schemes would be reversed.

More
Have students use paper wedges to generate a circle with 10 alternating black-and-white sections. Have them determine which angle openings they could use.

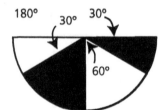

Answers for Student Page

C

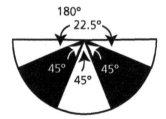

Reflection Designs

Each of the diagrams below has rotational symmetry. In this activity you will investigate different ways to generate the designs using the Reflect-It Hinged Mirror.

You will need several copies of Blackline Masters A–C.

- Position the Reflect-It Hinged Mirror to form a 90° angle. Place a blank 90° wedge in the mirror. Experiment to find a way to color the wedge to generate the diagram on the right.
- Repeat with your mirror at a 180° angle and a blank 180° wedge.

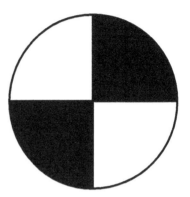

- Position your mirror at a 60° angle. Place a blank 60° wedge in the mirror. Generate the diagram at the right.
- Repeat with the mirror at a 120° angle and a blank 120° wedge.
- Try it again. This time position your mirror at a 180° angle and use a blank 180° wedge.

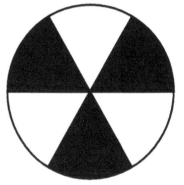

- Position your mirror to form a 45° angle and use a blank 45° wedge. Generate the diagram at the right.
- Repeat with your mirror at a 90° angle and a blank 90° wedge.
- Try it again. This time use a blank 180° wedge and position your mirror at a 180° angle.

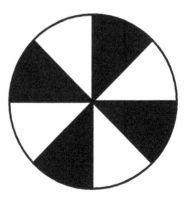

- There's more than one way to generate each design. For each different wedge you used, predict a different way to color the wedge that would generate the same design. Then test your prediction.

© ETA hand2mind®

Circle Challenges

Purpose
- To discover the meaning of factor

Materials
Several copies of Blackline Master A for each student, page 56
1 protractor for each student

Questions for Discussion
- Where did you draw the line segment on the blank wedge?
- Can you generate a circle with three equally spaced radii? with five equally spaced radii? with six equally spaced radii?
- How many different angle measures are possible between two radii in the first circle? in the second circle? in the third circle?
- What can you conclude about the three circles and your ability to duplicate the designs?

Help for You
Encourage students to experiment with placing the line segment in different positions on the wedge. Students may find it easier to manipulate a toothpick rather than drawing numerous segments. Running the line along one side of their mirror and pivoting it toward the other side of the mirror allows students to quickly investigate an endless number of possible positions.

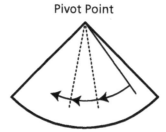
Pivot Point

Have students use a protractor to find the angle measure between different pairs of radii in each circle. Help students recall that a factor is a number that divides into another number without having a remainder.

Answers for Student Page
1. Three equally-spaced radii: 120°, 240°
   Five equally-spaced radii: 72°, 144°, 216°, 288°
   Six equally-spaced radii: 60°, 120°, 180°, 240°, 300°

2. Students will discover that the six-radii circle is the only pattern that can be duplicated using the 90° wedge. The 180° angle measure is a multiple of 90° and also occurs as an angle measure between any two of the six equally spaced radii on the circle. For the circles with three or five equally spaced radii, 90° is not a factor of any of the angle measures listed.

Activities for Reflect-It™ Hinged Mirror

Circle Challenges

In this activity, you will investigate what happens when you use the Reflect-It Hinged Mirror with circles.

You will need several copies of Blackline Master A.

■ Position the Reflect-It Hinged Mirror to form a 90° angle. Place a blank 90° paper wedge in the mirror.

■ Experiment to identify which of these circles you can generate with the mirror. For each, draw a line segment on a different wedge. The radii on the generated image should be equally spaced.

1. For each circle, what are the angle measures between any two radii in the circle?

2. What did you notice about your results for the three experiments? Write as many statements as you can.

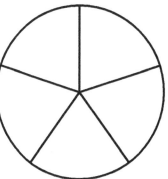

■ Predict what other numbers of radii you could generate with the 90° paper wedge. Then use your mirror to test your predictions.

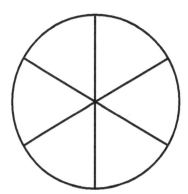

© ETA hand2mind®

Mini-Max Perimeters

Purpose
- To find minimum and maximum perimeters possible for figures with the same area

Materials
6 squares (1 cm × 1 cm) for each student
1 copy each of Blackline Masters E and F for each student, pages 60–61

Questions for Discussion
- What is the area of the reflected image?
- Does the area change as the squares are moved into different arrangements? Why or why not?
- When you try to find the minimum perimeter, why should the shaded square be covered?
- What is the minimum perimeter?
- What is the maximum perimeter?
- What is the relationship between the maximum perimeter when squares touch side to side and the maximum perimeter when squares can touch corner to corner as well as side to side? Explain.

Help for You
Initially, students will find it easier to experiment if they limit the squares to side-by-side arrangements with no holes in the generated image. In order to minimize the perimeter, students should place as many squares as possible side to side or against their mirror. In order to maximize the perimeter, students should place as few squares as possible side to side or against their mirror.

More
Students will need the same grid wedge, their mirror at the 90° angle opening, and 16 1 cm × 1 cm squares. Students start with 4 squares. Starting on the shaded box, students must place 4 squares on the grid wedge so that the generated image is a rectangle. Have them create a chart on which to record the number of squares used and the length, width, and area of the generated rectangle. Then have students rearrange the 4 squares to form as many other rectangles as possible, recording the same information for each.

Students can repeat the activity using 9 squares. Students can then analyze their data.

Answers for Student Page
See pages 67–68.

Mini-Max Perimeters

Can shapes have the same area but different perimeters? In this activity, you will find out.

You will need 1 copy of Blackline Masters E and F.

■ Position the Reflect-It Hinged Mirror to form a 90° angle.

■ Put the mirror on the grid wedge until the entire wedge is reflected in it.

■ Start with 3 squares. Find the generated image with the minimum perimeter possible.

Rules for Placing Squares
■ Place 1 square on the shaded corner square of the grid wedge.
■ Place the other squares on the wedge, 1 in each box, so that any 2 squares completely touch side to side. The generated image should not have any uncovered squares or "holes" in its interior.

1. Record how you placed the squares and what the minimum perimeter was.

■ Using the same 3 squares, experiment to find the generated image with the maximum perimeter possible.

2. Record how you placed the squares and what the maximum perimeter was.

■ Experiment with 4, 5, and 6 squares.

3. Record your results for minimum and maximum perimeter.

4. What patterns do you notice?

■ Now change the rule for placing squares. This time the squares can touch corner to corner. Experiment with 3, 4, 5, and 6 squares.

5. Record your results for minimum and maximum perimeter.

6. What patterns do you notice? Compare these results to the original results.

© ETA hand2mind®

Perimeter Puzzles

Purpose

To discover that geometric figures with the same perimeters can have different areas

Materials

6 squares (1 cm × 1 cm) for each student
1 copy each of Blackline Masters E and F, pages 60 and 61

Questions for Discussion

- How did you arrange the squares to generate a figure with the desired perimeter?
- How many squares did you use?
- What was the area of the generated image?
- What did you notice about the arrangements drawn on the recording sheet?

Help for You

Students may need two sessions to complete this activity.

Many students will discover that when a given number of squares are placed in different positions in the angle opening, identical images may appear. These two arrangements are themselves reflections—one working off the left side of the mirror and the other, off the right side of it.

More

Ask students to consider the same problem but allow them to create reflected images that contain holes. Have students select three, four, five, or six squares for the experiment and then arrange the squares on the grid side to side, so that the reflected image has a hole in its interior. The total perimeter, which is determined by adding together the perimeter of the inner border and the perimeter of the outer border, must remain constant.

Answers for Student Page

See pages 68–69.

Perimeter Puzzles

Can shapes have the same perimeter but different areas? In this activity, you will find out.

You will need a copy of Blackline Masters E and F.

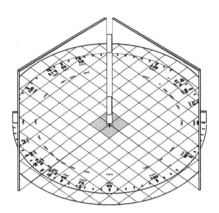

- Position the Reflect-It Hinged Mirror to form a 90° angle.
- Slide your mirror over the wedge as shown on the right.
- Generate a figure with a perimeter of 20.

Rules for Placing Squares
- Place 1 square on the shaded corner square of the grid wedge.
- Place the other squares on the wedge, 1 per box, so that any 2 squares touch side to side. The generated image should not have any holes or uncovered squares in its interior.

1. Record the area and how you positioned the squares.

- Experiment to find as many other ways as possible to generate figures with a perimeter of 20. Use 3, 4, 5, or 6 squares.

2. For each arrangement, record the area and how you positioned the squares.

- Try the experiment again. Generate figures with a perimeter of 24. Then generate figures with perimeters of 28, 32, and 36.

3. For each arrangement, record how you positioned the squares and the area and the perimeter of the generated figure.

4. What patterns do you notice? Write as many statements as you can.

© ETA hand2mind®

Blackline Master A: 90° and 180° Wedges

180°

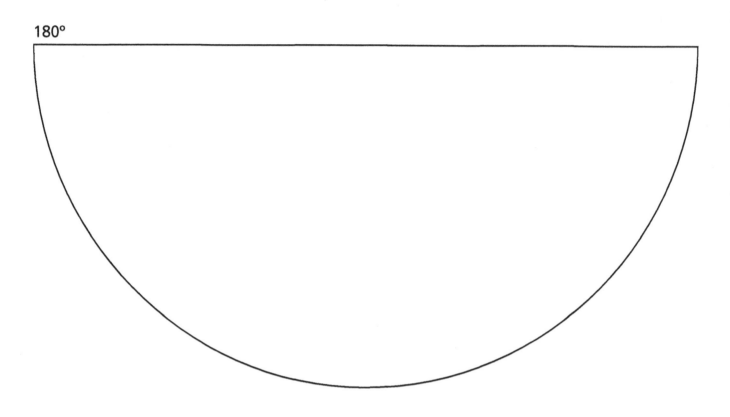

90°   90°

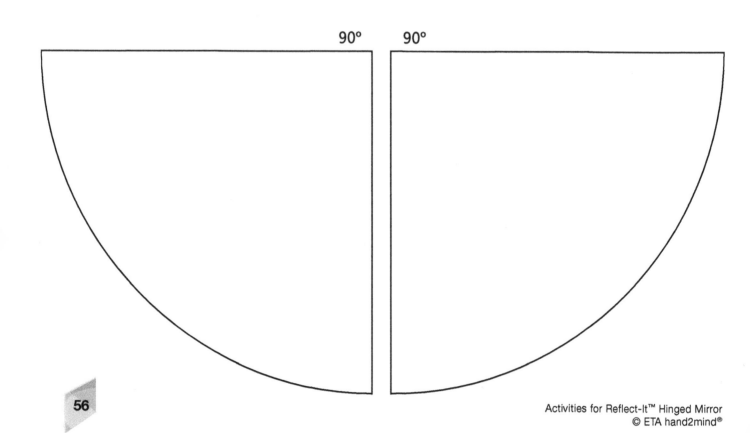

Activities for Reflect-It™ Hinged Mirror
© ETA hand2mind®

Blackline Master B: 60° and 120° Wedges

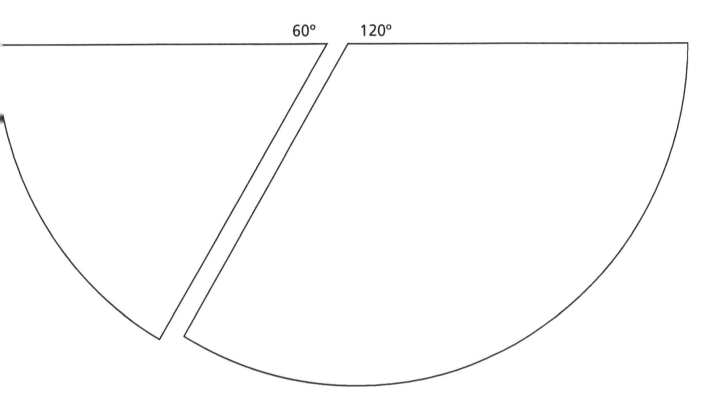

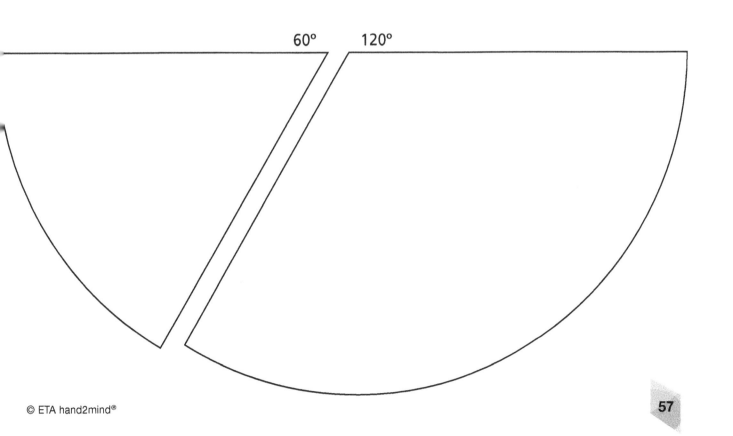

57

Blackline Master C: 20°, 30°, 36°, 40°, 45°, and 72° Wedges

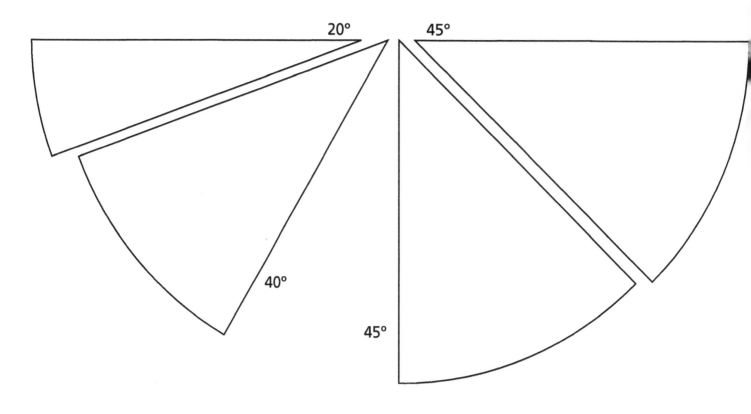

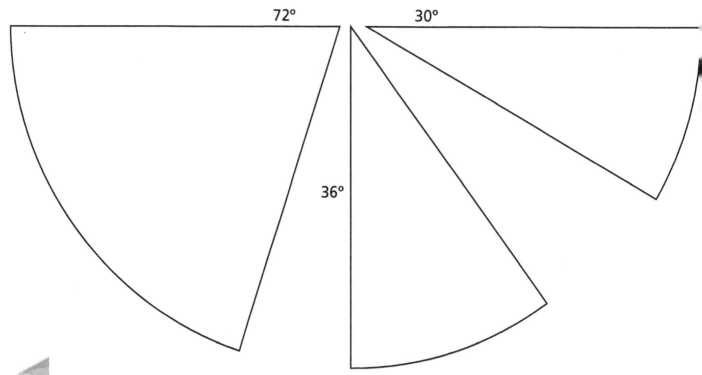

Activities for Reflect-It™ Hinged Mirror
© ETA hand2mind®

Blackline Master D: Stick Figure Wedge

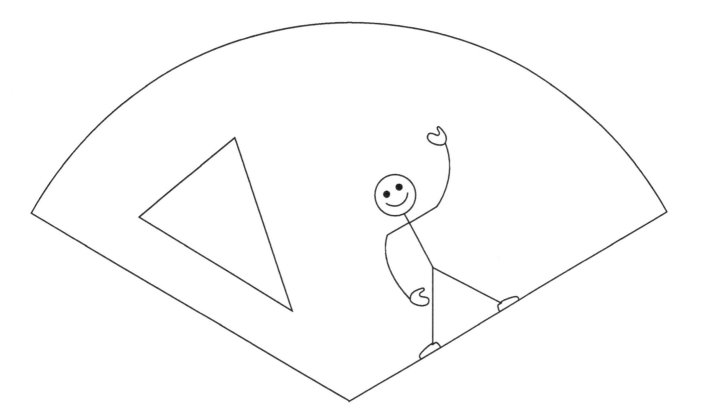

Blackline Master E: Grid Wedge

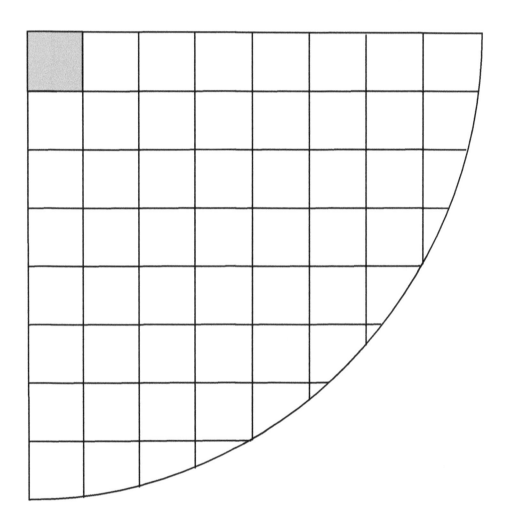

Blackline Master F: Grid Wedge Recording Sheet

© ETA hand2mind®

Blackline Master G: Road Signs

SIDE ROAD

ROAD NARROWS

CAMPGROUND

ROAD HAZARD

TRAFFIC CIRCLE

END OF ROAD

YIELD

CROSS ROAD

TWO-WAY TRAFFIC

STOP AHEAD

STOP

Blackline Master H: World Wedges

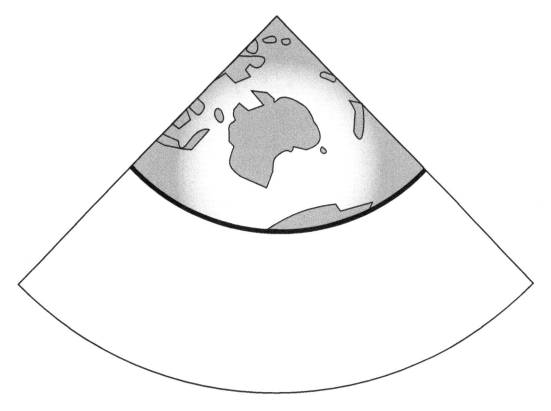

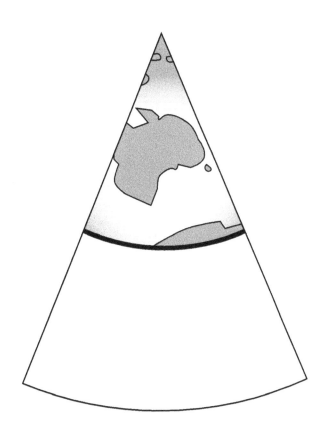

Blackline Master I: Reflect-It™

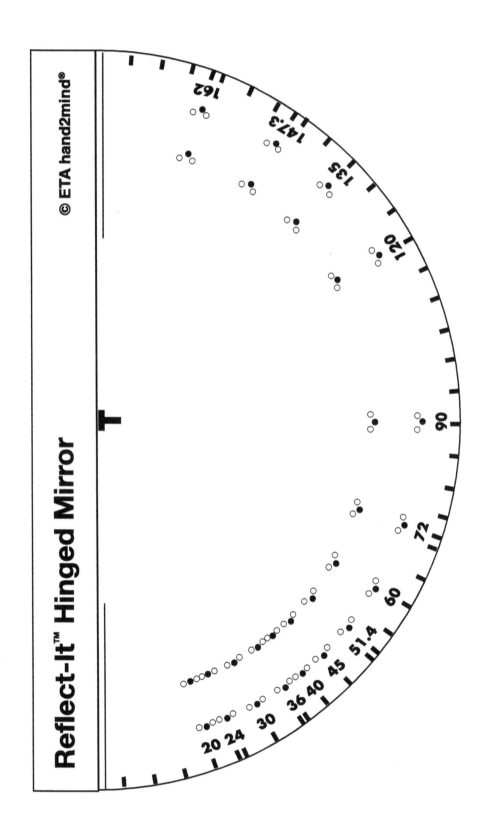

Blackline Master J: Cosmic Experience

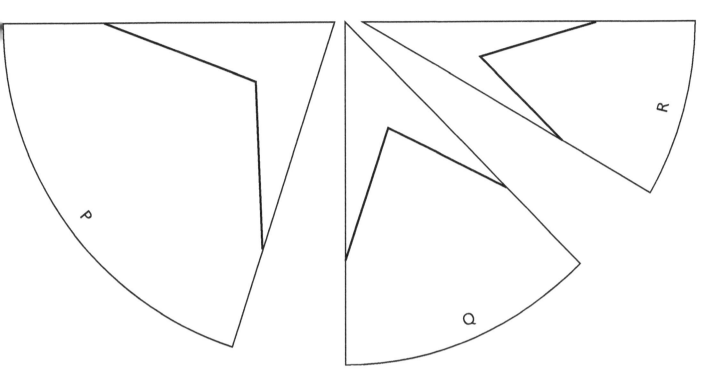

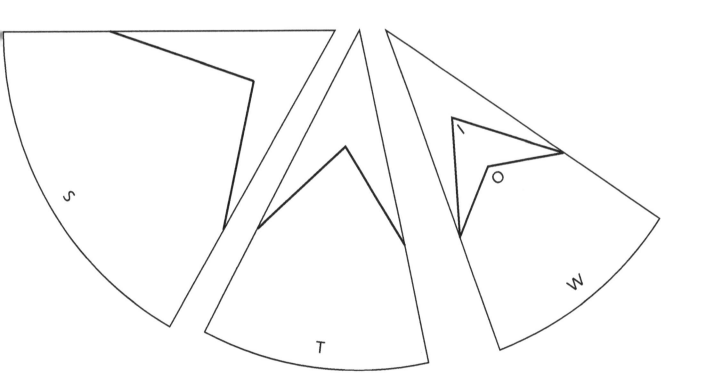

65

Blackline Master K: Blank Circle

Additional Answers

Name That Polygon (pages 40–41)

Angle Opening in Degrees	Number of Sides in Polygon	Name of Polygon	Measure of Each Angle in the Polygon
120°	3	triangle	60°
90°	4	square	90°
72°	5	pentagon	108°
60°	6	hexagon	120°
45°	8	octagon	135°
40°	9	nonagon	140°
36°	10	decagon	144°
30°	12	dodecagon	150°
20°	18	18-gon	160°

Mini-Max Perimeters (pages 52–53)

1.–3.

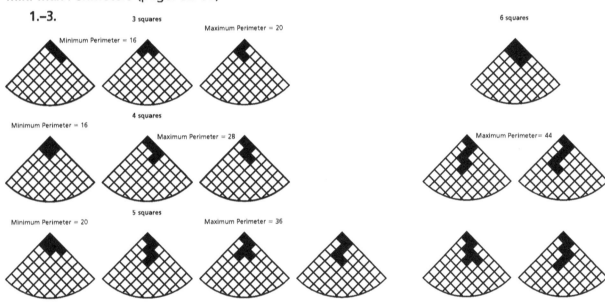

4. Students may notice the following patterns:
   For each set of squares, the area remains constant.
   Figures with the same area can have different perimeters.
   Each time a square is added, the maximum perimeter increases by 8.
   All the perimeters are even numbers.

5. Minimum perimeters are same as for 1.–3.

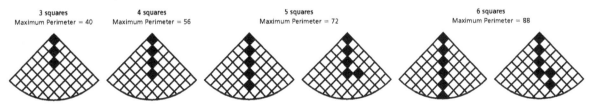

© ETA hand2mind®

67

Additional Answers

Mini-Max Perimeters (continued)

6. Some patterns students may notice are:
   Each time a square is added, the maximum perimeter increased by 16.
   The maximum perimeters are twice those made with the original rule.
   The minimum perimeters are the same with both rules.

Perimeter Puzzles (pages 54–55)

3.

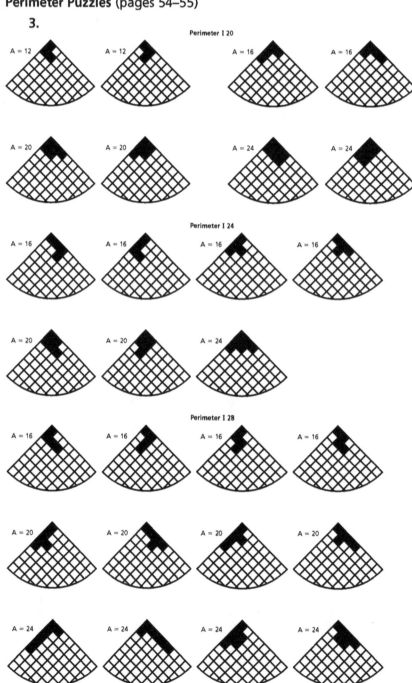

Additional Answers

Perimeter Puzzles (continued)

Note: For each answer shown, the same arrangement working off the opposite mirror edge will generate an identical image.

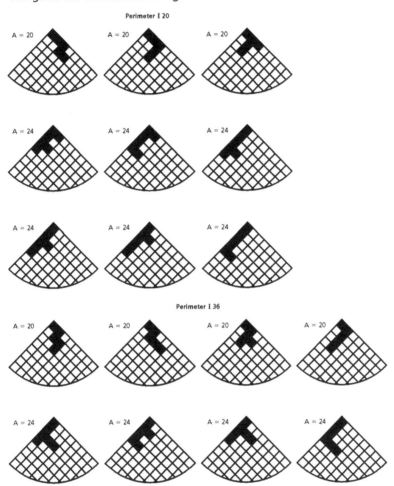

Glossary

Diagonal line symmetry
A figure has diagonal line symmetry if it can be folded along a line running at a 45° angle with the horizontal so that the two halves are exact mirror images. If the figure is folded along the line of symmetry, the two halves will coincide. (page 22)

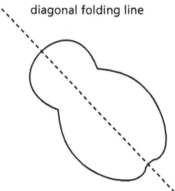
diagonal folding line

Flip
A reflection. (page 30)

Generated images
Multiple images created by the Reflect-It Hinged Mirror, including the original image. (page 14)

Horizontal line symmetry
A figure has horizontal line symmetry if it can be folded along a horizontal line so that the upper half is the exact mirror image of the lower half. If the figure is folded along the line of symmetry, the two halves will coincide. (page 20)

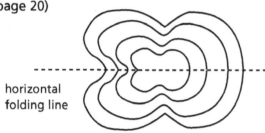
horizontal folding line

Indirect variation
This occurs when two quantities are related in such a way that as one quantity increases, the other decreases; or, as one quantity decreases, the other increases. The product of two quantities that vary indirectly is a constant value. (page 14)

Line symmetry
A figure has line symmetry if it can be folded along a line so that one half is the exact mirror image of the other half. If the figure is folded along the line of symmetry, the two halves will coincide. (page 20)

Reflection
The mirror image of an object; a flip. (page 8)

Reproduction
An exact copy, or duplicate, of another object. (page 9)

Rotational symmetry
A figure has rotational symmetry if, after a specific number of rotations of less than 360° around an internal pivotal point, the figure returns to its original position. (page 32)

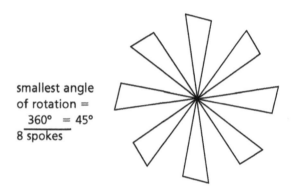
smallest angle of rotation = $\frac{360°}{8 \text{ spokes}}$ = 45°

Vertical line symmetry
A figure has vertical line symmetry if it can be folded along a vertical line so that the left half is the exact mirror image of the right half. If the figure is folded along the line of symmetry, the two halves will coincide. (page 16)

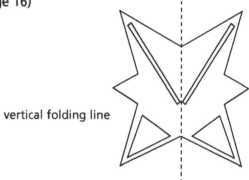
vertical folding line

Activities for Reflect-It™ Hinged Mirror
© ETA hand2mind®